A SWARM OF BEES

BY REBECCA STORM

CONTENTS

HONEYBEES	4
PARTS OF A BEE	6
LIVING IN THE HIVE	8
QUEEN BEE	10
CELLS OF LIFE	12
HARD WORKERS	14
FLOWER HUNT	16
THE WAGGLE DANCE	18
BUSY AS A BEE	20
HUMANS AND HONEY BEES	22
ALL SORTS OF BEES	24
WANNABEES	26
FUN BEE FACTS	28
GLOSSARY	30
INDEX	32

Copyright © 2025 Hungry Tomato Ltd

First published in 2025 by Hungry Tomato Ltd
F15, Old Bakery Studios, Blewetts Wharf, Malpas Road, Truro, Cornwall,
TR1 1QH, UK.

No part of this publication may be reproduced, stored in a retrieval system, or transmitted in any form or by any means, electronic, mechanical, photocopying, recording, or otherwise, without prior written permission of the copyright owner.

A CIP catalogue record for this book is available from the British Library.

ISBN 9781835694206

Printed in China

Discover more at
www.hungrytomato.com

DISCLAIMER:
Insects are fascinating, but best to stay away! Don't touch or handle them – some insects can sting or get aggressive when they feel threatened.

Picture credits:
Abbreviations: m-middle, t-top, l-left, r-right, bg-background.

FLPA Images: 10m, 11br. Science photo library: 15b. Ticktock Image Archive: 27bl, 27tr. Shutterstock: 9br; Aleksandr Rybalko 21b; Anthony King Nature 11t; bamgraphy 13t; BushAlex 26tl; Cherryblossom 9ml; Cornel Constantin 3bl, 17t; Daniel Prudek 6-7bg, 31; dsangid 15tr; fevzie 1bg, 12b; Hollygraphic 19b; Jeff Holcombe 17b; jordan roper 13br; J_Frefz 25tl; kuttelvaserova Stuchelova 9bl, 28bl; Lois GoBe 16m; Malaha 24tl; Maurice Lesca 25bl; meggedyannephotography 4tl, 18b, 29tr; Minko Peev 20b; Mykhailo Pavlenko 7tr; New Africa 22b, 23br; nicemyphoto 4br; Orest Malanchu 9tl; OlgaBerlet 8b; Ozgur kerem Bulur 26br; Perpis 27tr; Photografiero 21tr; SanderMeertinsPhotography 5tr; Sener dagasan 29ml; Seyfettin Karagunduz 25b; tanabora 24br; rayints 28mr; vetlas 13bl.

Every effort has been made to trace the copyright holders, and we apologise in advance for any unintentional omissions. We would be pleased to insert the appropriate acknowledgements in any subsequent edition of this publication.

Words in **BOLD** can be found in the glossary.

HONEYBEES

Honeybees are medium-sized winged **insects**. They look fluffy, but can deliver a painful sting too! Honeybees make a distinctive buzzing sound when they fly. They are most well-known for making **honey**!

HOW DO THEY LIVE?

Honeybees belong to a group of insects known as social insects because they live in large family groups called a **colony**. Most of the honeybees in a colony are **workers** that collect **nectar** and **pollen** from flowers.

A colony of bees in a **hive**.

WHERE DO THEY LIVE?

Honeybees do not like very cold conditions, and they do not like it very hot and dry either. Honeybees will live wherever there are plenty of flowers, in the warmer months of the year.

WHAT DO THEY EAT?

Honeybees eat nectar and pollen from flowers, and **resin** from trees. They have fantastic flight control, and can check a flower for pollen while hovering in the air!

IT'S A BUGS WORLD

Insects belong to a group of bugs known as **arthropods**. They have an outer "skin" called an **exoskeleton** that protects their bodies, instead of an inner **skeleton** made of bones. All adult insects have six legs, and most also have at least one pair of wings for flying, although some have two pairs.

PARTS OF A BEE

A honeybee worker has six legs and two pairs of flying wings. A layer of thin hairs gives this bug a fuzzy appearance. Its body is divided into three parts – head, **thorax**, and **abdomen**.

The abdomen is the largest part of the honeybee's body. It contains the **digestive system**, and other important **organs**.

Inside the upper abdomen, a honeybee has a special organ called the "**honey** stomach". Nectar is stored here, instead of passing through the digestive system.

SIX LEGS

Bees and other insects are sometimes called "hexapods" because they all have six legs ("hex" means "six" in Greek). This can be a bit confusing – all insects are hexapods, but not all hexapods are insects!

Honeybees are insects, so they have six legs.

The thorax is the middle part of the body, attached with legs and wings.

The head is equipped with antennae, eyes, brain, and mouth.

LIVING IN THE HIVE

A honeybee colony is a huge family with thousands of closely related members. A colony of honeybees make themselves a nest called a hive. The hive is a very busy and highly organised place!

Honeybees often set up colonies in hollow trees. If no hollow trees are available, they may build a nest on a branch. Honeybees will also build hives under roofs, on the ground, underground, and in caves.

Hives built on rocks

Every honeybee colony contains three types of honeybees – workers, **drones**, and a **queen**.

Most of the honeybees are female workers. The queen bee is bigger than the workers, and does not leave the hive. Drones are male honeybees – they are smaller than the queen, but bigger than workers.

Queen bee

Worker bee

Drone bee

TREASURE HOUSE

A hive not only provides a home, it's also the colony's treasure house. As well as the queen bee and her growing **larvae**, the hive also contains the colony's precious store of food! To protect it, worker bees often make a wall of mud around the entire hive.

QUEEN BEE

Queen bees are the largest of all honeybees. A queen bee mates just once, early on in adult life. The queen mates with drones from nearby hives during one mating flight. The queen then creates a new colony.

Fed and cared for by worker bees, the queen lays thousands of eggs a day for the rest of her life! She never leaves the hive again, unless there's an emergency. Most eggs hatch and grow into worker bees, so that they can keep the hive safe.

Worker honeybees surrounding a queen honey bee (marked with a pink dot).

The queen sometimes lays eggs that hatch into drones. A queen honeybee will only lay "queen eggs" when the colony needs a new queen.

KILLER QUEENS

To make a new queen bee, worker bees choose 10-20 young female larvae and feed them a special food called royal jelly. This food helps the larvae grow bigger and faster. It also turns them into queens, giving them the ability to lay eggs, which worker bees can't do!

CELLS OF LIFE

The inside of a hive is an amazing example of insect architecture and engineering. The queen lays each egg into an individual cell that has been made by the workers from beeswax.

Each cell is perfectly designed to fit together. After the egg has been laid, the cells are sealed shut. Inside the cell, the egg hatches into a larva. The larvae are fed and cared for by workers that regularly open the cells, feed the larvae, and then reseal the cells.

Young bee being fed in its cell by a worker bee

When a larva reaches full size, it **pupates** and creates an outer casing around itself. At this stage in life, a bee is known as a pupa. Inside the casing, the pupa changes into the body shape of an adult.

A close-up of a pupa honeybee in its casing

INSECT DEVELOPMENT

Insects develop from eggs in two different ways. Some insects, including bees, hatch into larvae, which look very different from the adults. Other insects, such as stink bugs, hatch into **nymphs**, which already have the body shape of an adult.

Honeybee larvae

Stink bug nymphs hatching

HARD WORKERS

After young worker bees emerge from their cells, they spend about a month inside the hive. At first, their duties keep them among the cells.

Young workers keep the cells in good condition, and also help to make new ones. Worker bees can make small amounts of wax in special **glands** on the sides of their bodies.

A young worker bee repairs cells in the hive

They use their legs to scrape this wax into small balls that can be carried around the hive. Young workers chew the wax to make it sticky, then use their mouths to shape it and press it into place.

The young then progress to guard duty. Their job is to keep out any predators! They also flutter their wings at the entrance to help create air currents that keep the hive cool.

Guard bees attack an intruder drone bee

STING SACRIFICE

When a worker honeybee stings a predator, it sacrifices its life. A bee's sting has **barbs** that catch onto skin and hold the sting firmly in place while poison is injected. When the worker honeybee moves away, the sting tears from its body and is left behind, and the bee dies.

The used sting of a worker honeybee

FLOWER HUNT

Honeybees really like plants with lots of flowers – the more flowers, the more pollen and nectar! Plants that have flowers really like honeybees, and they do their best to attract them.

Flowering plants produce pollen so that one flower can help another to produce seeds. Honeybees collect pollen and take it away, accidentally helping the plants by carrying it from flower to flower. This process is called pollination.

SWEET TASTE

Many plants produce small amounts of sweet, sugary nectar, a very high-energy food designed to be especially appealing to insects, especially bees. Nectar is the favourite food of honeybees!

Bees have excellent eyesight, although they can only see in **UV** light. Flower petals often have patterns that are designed to attract honeybees, but are invisible to other animals, such as humans!

Honeybee collecting nectar from the flowers

THE WAGGLE DANCE

Each morning, scout bees are sent out to search for the best place to find pollen and nectar. Any worker can be a scout bee, but usually it is the older, more experienced bees that are chosen.

The scout bees fly long distances in all directions from the hive to investigate new flowers. When each scout returns to the hive, it reports what it has found using the waggle dance!

A female honeybee performing a waggle dance to tell other bees where to find pollen!

Returning scout bees perform a dance at the hive entrance. The main part of the dance highlights the direction to go to find food. The different ways in which they dance provides information to other bees about the distance, amount, and quality of the pollen and nectar.

HONEYBEE NAVIGATION

Honeybees find their way around by using the sun. Not only do bees use the direction of the sun, but also the height of the sun in the sky, and the strength of sunlight, as it varies during the day. This allows honeybees to know exactly where they are, even when they are far from the hive.

BUSY AS A BEE

For the rest of their short lives, worker bees collect pollen and nectar every day. This mainly goes towards feeding bees in the hive – the queen, young workers, and drones – then the rest of the food goes into long-term storage.

Worker honeybees collect pollen with their mouths and front legs, then carry it back in special pollen baskets on their back legs. Any nectar that is collected is carried inside a worker's honey stomach.

A honeybee with pollen collected on its legs

Any food that is not eaten is stored in empty egg cells that have been cleaned and resealed.

As well as pollen and nectar, honeybees also collect water and tree resin, which they use for sealing the ends of the cells.

Honeybee drinking water

HONEYCOMBS

Inside the individual wax cells, the nectar and pollen mixture slowly dries out into a sticky substance called honey. A collection of honey cells is called a **comb**, otherwise known as honeycomb.

A worker bee in a honeycomb

HUMANS AND HONEYBEES

Many animals enjoy the taste of honey, and some even enjoy the taste of bees! Humans like the taste of honey so much that they turned honeybees into farm animals.

A beehive can be made by humans from mud, straw, or wood. If the colony is given sugar syrup when there is no pollen or nectar available, all the stored honey can be taken without harming the bees. Honeybees are kept in most countries.

A beekeeper looking after some bees

Each bee sting contains a very small amount of mild poison, and very few people are badly affected by a single sting. Lots of stings can harm a person, so beekeepers are always careful! They often wear white suits to protect themselves.

A beekeeper in a white protective suit

BY CANDLELIGHT

Beeswax provides people with another benefit from honey gathering or beekeeping. For thousands of years, the finest candles have been made from beeswax. Today, beeswax is still used for the most expensive and luxurious candles.

ALL SORTS OF BEES

There are many species of bees in the world. Most live differently to honeybees – some live in smaller colonies, and some live alone. Most bees eat nectar and pollen, but no other species build such big stores of honey as honeybees!

BUMBLEBEES

Most bumblebees are much bigger, hairier, and noisier than honeybees. Bumblebees have typical warning colours: black and yellow or black and red. They live in far smaller colonies, and unlike honeybees, bumblebees are capable of stinging more than once.

CARPENTER BEES

A carpenter bee is about the size of a bumblebee. The female carpenter bee uses its sting to cut out narrow holes in pieces of dead wood. At the bottom of each hole, the bee lays a few eggs.

CUCKOO BEES

These solitary bees are **parasites** on other bees. Like cuckoo birds, cuckoo bees do not build their own nests. Instead, female cuckoo bees search out the nest of another species of solitary bee and steal it. They sting all the eggs and larvae to death before cleaning out the cells so that they can lay their own eggs.

LEAF-CUTTER BEES

Leaf-cutter bees live by themselves. They have powerful **jaws** that are used for digging into soil and for cutting leaves. After digging a tunnel, the female cuts up leaves from nearby plants. Each piece is rolled into a tube and placed into the tunnel. An egg is laid in this "leaf cell".

WANNABEES

Identifying different bees is made even more difficult by the fact that lots of other insects want to look like bees! Animals that pretend to be bees are known as bee-mimics.

DRONE FLY

The drone fly belongs to a group of flying insects known as hoverflies, many of which are bee or wasp mimics. The drone fly looks just like a harmless honeybee drone, and it tries to sneak into hives to feed on honeybees' stored pollen.

GREATER BEE FLY

The greater bee fly is a fly that looks just like some bee species, except that is has a long mouthpart. It uses this to feed on nectar. Females scatter their eggs while flying around. When the eggs hatch on the ground, the flies seek out the nests of bees and eat their larvae.

VELVET ANT

This bug has the hairy appearance of a bee, but despite its name, is actually a species of wasp! Velvet ants are very aggressive. The female lays her eggs inside the bodies of bee larvae and pupae. When the velvet ant larvae hatch, they begin eating the developing bees!

BUMBLE FLY

The bumble fly is one of the hoverflies that use their appearance as a means of protection. It feeds on nectar and pollen and relies on predators mistaking it for a real bumblebee. The bumble fly, like other hoverflies, has no sting.

FUN BEE FACTS

There's so much more to know about bees! Delve into some fantastic facts about these clever, little insects.

BEES ORIGINATED IN...
Tropical Africa, and then spread all around the world! Today, they can be found everywhere, aside from Antarctica.

HONEYBEES HAVE BEEN USED SINCE...
the age of the pyramids, around 5,000 years ago!

EACH COLONY CAN HAVE...
tens of thousands of bees inside. Some colonies contain as many as 80,000 bees!

THE FIRST BEE FOSSILS...
are about 40 million years old, and are almost identical to modern bees.

ROYAL JELLY IS...
bee milk produced by worker bees after they have eaten pollen.

SCIENTIST VON FRISCH DISCOVERED...
honeybees communicate through the language of dance.

BEES PRODUCE ROYAL JELLY...
that is constantly fed to the queen, as well as to her larvae after they hatch.

A QUEEN LAYS...
around 2,000 eggs a day!

HONEY HAS BEEN...
used to heal injuries for many years.

GLOSSARY

Abdomen – the largest part of an insect's three-part body; the abdomen contains most of the important organs.

Arthropod – any small bug that has jointed legs – insects and spiders are arthropods.

Barbs – sharp points facing the opposite direction on a weapon or tool that stops it being taken out of the victim.

Beeswax – a sticky, solid substance produced by honeybees and used for building cells.

Cell – a hollow, six-sided structure made by honeybees to raise their young and to store honey.

Colony – a group of insects, or other living things, which live very closely together.

Comb – a collection of cells built side-by-side inside a hive.

Digestive system – the organs that are used to process food.

Drones – male honeybees; drones are larger than worker bees, but do not have a sting.

Exoskeleton – a hard outer covering that protects and supports the bodies of some animals.

Glands – a part of an animal's body that produces particular substances.

Hive – the nest made by a colony of honeybees.

Honey – sweet, syrupy substance produced by honeybees from pollen and nectar.

Insect – a type of very small animal with six legs, a body divided into three parts, and usually two pairs of wings.

Jaws – hinged structures around the mouth that allow some animals to bite and chew.

Larvae – wormlike creatures that are at the juvenile (young) stage in the life cycle of many insects.

Mate – one of a pair of animals that live or have babies together.

Nectar – a sweet, sugary substance produced by flowering plants and used by honeybees to make honey.

Nymph – the juvenile (young) stage in the life cycle of insects that do not produce larvae.

Organs – parts of an animal's body that performs a particular task.

Parasites – living things that live or feed on or in the body of another living thing.

Pollen – tiny grains that are produced by flowers in order to fertilise other flowers.

Pupates – an insect larva in the process of turning into an adult.

Queen – the largest honeybee in a colony; the queen is the only female bee that can lay eggs.

Resin – a thick, sticky substance that comes from trees.

Skeleton – an internal structure of bones that supports the bodies of large animals such as mammals, reptiles, and fish.

Thorax – the middle part of an insect's body where the legs are attached.

UV – also known as ultraviolet, UV is light that cannot be detected by human eyes, but can be seen by honeybees.

Worker – female honeybees – nearly all the honeybees in a hive are workers.

INDEX

A
abdomen 6, 30
adult insects 5
antennae 7
arthropods 5, 30

B
barbs 15, 30
bee milk 29
bee mimics 26–27
beekeepers 22-23
bees, types 24–25
beeswax 12, 23, 30
   (see also: wax)
bumble fly 27
bumblebees 24

C
candles 23
capped cells 12–13
carpenter bees 24
cells 12–13, 30
colonies 30
combs 21, 30
cuckoo bees 25

D
dance 18–19, 29, 31
digestive system 7, 30
drone fly 26
drones 9, 10–11, 28–30

E
eating habits 5
   (see also food)
eggs 10-11, 12-13
exoskeleton 5, 30

F
flies 26–7
flight control 5
flowers 5, 16–19
food 16-17
flowers 5, 16–19
fossils 28

G
glands 14, 29, 30
greater bee fly 26
guard bees 14–15

H
hats, beekeepers 23
head 6–7, 15
hexapods 7
hives 5, 8–9, 11–15, 30
honey 22, 30
honey-stomach 7, 20–21
honeycombs 21
hoverflies 26–27
humans 22–23, 29

I
insects 4-5, 6-7, 8-9, 10-11, 12-13, 14-15, 16-17, 18-19, 20-21, 22-23, 24-25, 26-27

J
jaws 25, 30

L
larvae 12–13, 28–29, 30
leaf-cutter bees 25
legs 5, 6-7, 14, 20

M
mating flights 9, 10, 29
mouth parts 20, 27
mud casing, hives 9

N
navigation 19
nectar 5, 7, 17–21, 31
nests (see hives)
nymphs 13, 31

O
organs

P
parasites 25, 27, 31
plants (see flowers)
poison 23
pollen 5, 16–17, 18–19, 20–21, 31
pollination 17
pupa 11, 13, 28, 31

Q
queen's control 10–11
queen eggs 11
queens 9–12, 24, 28–29, 31

R
repairs, hives 14–15
resin 5, 21
royal jelly 28, 29

S
scout bees 18–19
skeleton 5, 31
social insects 4
stings 29
sun navigation 19

T
thorax 6–7, 31

U
ultra-violet (UV) light 17, 31

V
velvet ant 27
von Frisch, Karl 29

W
waggle dance 18–19, 31
wasps 27
water 21
wax 12, 14, 21, 23, 30
wings 5, 15, 19
workers 4, 6, 8, 9, 12, 14-15, 30-31